José Heriberto Simental Vázquez

Automated Baseball Pitcher

AF524734

José Heriberto Simental Vázquez

Automated Baseball Pitcher

Design and development of an automated baseball pitcher

ScienciaScripts

Imprint
Any brand names and product names mentioned in this book are subject to trademark, brand or patent protection and are trademarks or registered trademarks of their respective holders. The use of brand names, product names, common names, trade names, product descriptions etc. even without a particular marking in this work is in no way to be construed to mean that such names may be regarded as unrestricted in respect of trademark and brand protection legislation and could thus be used by anyone.

Cover image: www.ingimage.com

This book is a translation from the original published under ISBN 978-613-9-41012-5.

Publisher:
Sciencia Scripts
is a trademark of
Dodo Books Indian Ocean Ltd. and OmniScriptum S.R.L publishing group

120 High Road, East Finchley, London, N2 9ED, United Kingdom
Str. Armeneasca 28/1, office 1, Chisinau MD-2012, Republic of Moldova, Europe
Printed at: see last page
ISBN: 978-620-8-07660-3

Copyright © José Heriberto Simental Vázquez
Copyright © 2024 Dodo Books Indian Ocean Ltd. and OmniScriptum S.R.L publishing group

Book: "Design and Development of an Automated Baseball Pitcher".

Authors:

Alfonso Sierra Chacón
Efren Armando Lujan Sandoval
José Heriberto Simental Vázquez

Co-authors:

Oscar Iván Soto Guaderrama
Yolanda Isabel Villalobos Morales

INDEX

CHAPTER I
INTRODUCTION

1.1 Background

Today there are different prototypes with various operating systems inball dispensing machines, but the first pitching machine was invented by Charles Hinton in the mid 1890s.

The need for a pitching machine is to avoid constant injuries in baseball training at amateur, semi-professional and professional levels, as it is required to simulate match pitching conditions for all batters on the team, placing excessive muscle loads on the pitchers (usually fewer players than batters) causing muscle and joint injuries.

1.2 Problem statement

Modern society increasingly aims at the technification of sport, as in the caseof baseball schools and professional clubs that are dedicated to the professional development of players. In almost all professional baseball schools, technological equipment such as a pitching machine is used in the training and preparation of a high-performance hitter or catcher.

The Instituto Tecnológico de Ciudad Juárez does not have a ball machine for the baseball team.

1.3 Objectives

1.3.1 General Objective

Apply mechanical engineering knowledge to design an automated baseball pitching machine to reduce manufacturing cost vs. market cost".

1.3.2 Specific Objectives

a) Material allocation for the design of an automated ball throwing machine.

b) Use Solid Works software to design fixtures and mechanical elements for the operation of the ball machine.

c) Make use of engineering skills to control the launch, direction and power of the ball.

d) Design a functional machine reducing costs against the market.

e) Creation of drawings of fixtures and assemblies with BOM of materials.

1.4 Justification

Multitasking machines are constantly evolving. Since the inception of the concept, multitasking machine configurations have progressed from the first systems that combined simple operations, this time the baseball pitching machine will improve the batter's or catcher's dexterity and reflexes.

A professional ball throwing machine on the market costs approximately 26,000 Mexican pesos, the purpose is to provide the institute with a functional machine at an affordable cost, applying the knowledge of mechanical engineering to develop an optimal and functional design of the equipment.

1.5 Assumption

Design a pitching machine using commercial off-the-shelf components and optimise it using the Solidworks programme with fixtures for possible fabrication and assembly, then build a prototype that will fit the budget and contribute to the training of the batter or catcher.

The aim is to test the effectiveness of a machine by implementing it as a support in baseball training, in order to experience more favourable results.

CHAPTER II
BACKGROUND

2.1 Theoretical framework

2.1.1 Launching Machines

There is a wide variety of pitching machines on the market for sports such as football, basketball, tennis, baseball, etc. These machines contribute to the training of athletes in the different disciplines and have different operating and functioning systems. It is necessary to carry out a bibliographic study of the propulsion systems of the object to be thrown, distinguishing the characteristics of each one so that these previous investigations serve to find a relationship between concepts, theories and to locate different perspectives in order to apply them correctly.

2.1.2 Propulsion Systems

There are several systems for propulsion of spherical objects, among which we have:

- By air pressure.
- By compression and decompression of springs.
- By catapult.
- By rotating rollers.

It is necessary to review each of these systems for the different characteristics depending on the size of the object to be launched, the speed and time of continuous launch, the power supply of the system, as well as looking for a functional design, adequate manufacturing cost and feasibility of construction.

2.1.3 Air pressure propulsion systems

In this system one side is open and the other end is sealed, the element to be thrown is placed inside the cylinder, which is divided into two parts: in the first one the tennis ball is placed and in the second one it contains the compressed air that reaches a desired pressure using an air compressor, the two parts are connected by a quick opening valve that allows the passage of air from the compressor which gives the impulse to the baseball at high speeds. One of the big disadvantages of this system is the need for an air compressor and that requires a lot of time to recharge and transport it, also the continuous launch of balls is slow because the same exit hole of the balls must enter the next this makes the time between pitches is very long and not so advisable if you need continuous shots.

Baseball and softball pitching machine.

Note: The figure represents a pitching machine for baseballs and softballs.

2.1.4 Propulsion systems

Catapult type mechanism This machine has a mechanical system consisting of an electric motor installed with a reducer that transmits the movement by means of a chain making the catalina rotate on the upper axis, which has at each end a mechanism that allows the helical spring to store elastic energy, as can be seen in the image, the arm receives this energy that allows the ball to fall and start to rotate slowly, then it reaches the point where this arm acts as a catapult when the spring is released and launches the ball at high speeds. This machine is used in cricket training, which with the time of use requires constant maintenance of the mechanical part, especially due to the wear produced in the helical spring, in this type of mechanism the time of launching and waiting is prolonged.

2.1.5 Rotary roller propulsion systems

This type of system is most commonly used for football throwing machines. Within this group there are a few propulsion options:

- Single rotating roller.
- Two rotating rollers.

- Rotating rollers with belts.

2.1.6 Single-roller propulsion systems

The system shown in Figure 3 has an electric motor attached to a roller which rotates at a given angular velocity. The baseball enters the feeder and is propelled by the speed and pressure exerted on it by the roller. This type of machine is ideal for throwing small objects, due to its shape and layout, it has a small exit hole where the ball is thrown at high speeds. In terms of design and construction this machine provides certain ergonomic features that allow the user to have an easy experience when using this ball launcher.

2.1.7 Two-roller propulsion systems

The principle of operation of this type of system is the same as that of a single roller, the difference lies in the fact that by using two rollers, curved throws can be made, by controlling the speed of each of the rollers, one of them turning faster than the other will allow us to propel the ball along a curved trajectory, which the single roller system did not allow, but in the same way if a straight throw is required, both rollers must be turning at the same speed.

2.1.8 Rotating roller drive system with belts

The type of system shown in Figure 5 has an operating principle very similar to that of the two-roller system, with drums on which the two belts are attached to the drive rollers, this system has a larger surface area in contact with the ball, which allows a higher exit speed. Although the trajectory speed of the ball is higher than that of other systems, the disadvantage is the construction, which is very robust and mechanically more difficult than the previous systems.

2.1.9 Machines on the market today.

There are several machines on the market with different ratios, costs and capacities, this project focuses on professional machines with similarities to a 90 km/hr launch.

Table 1 Brands and costs

Machine on the market	Cost	Launch speed	Size
PowerNet	$ 8,931	40 - 90 MPH	25ft x 1ft x 3 ft
Heater Pro	$ 19,850	70 MPH	14ft x 1ft x 3ft
Hack Attack	$ 15,790	100 MPH	5ft x 4 ft x 4 ft
Sports Attack	$ 59,995	70MPH	8ft x 4 ft x 2 ft

2.2 Contextual framework

This chapter describes the institution at the local level, the Instituto Tecnológico de Ciudad Juárez (also known by its acronym ITCJ), is a public institution of higher education located in Ciudad Juárez, Chihuahua. It is part of the Tecnológico Nacional de México (TecNM) and the Mexican Ministry of Public Education.

For 55 years, the Instituto Tecnológico de Ciudad Juárez has been responsible for the technical and humanistic education of professionals who currently occupy an important place in the labour sector of the border.

Serving as one of the main educational centres at the professional level, the campus was officially recognised as the Instituto Tecnológico de Ciudad Juárez on 3 October 1964.

After having functioned in 1935 as Técnica Industrial 5 at the initiative of Professor Alberto Álvarez, it later changed its name to Escuela de Enseñanzas Especiales 21, which was located in the monument to Benito Juárez.

Norberto López, head of the ITCJ's Communication and Dissemination department, commented that this came about after Adolfo López Mateos, President of Mexico at the time, made a commitment to build a decent institute in 1960.

"At that time the city was beginning to ask for technicians who were a little more specialised in the simple industry that existed here, there were around 30,000 inhabitants," he said.

It was not until 1964 that the Instituto Tecnológico Regional de Ciudad Juárez (ITRCJ) was inaugurated, a name that was given to it to cover the different municipalities of the state, such as Cuauhtémoc, Nuevo Casas Grandes, among others.

Later, a juvenile prison was converted into the building that today prepares more than 7,000 students in the 10 engineering programmes offered by the Institute, including electromechanics, mechanics, mechatronics, logistics and business management, in addition to two academic programmes in public accounting and business administration, as well as three master's degrees and a doctorate.

Until the 80's the regional part was suppressed to finally become the Instituto Tecnológico de Ciudad Juárez, where there are currently 46 classrooms, so that the university came to meet the demand of all the people of Juarez interested in the industrial sector, when there was no great boom in the city.

In 2014, the 254 campuses in the country were brought together to be known with a new identity 'Tecnológico Nacional de México', considering each institute as a campus.

Figure 1 Tecnológico Nacional de México, Campus Ciudad Juárez.

2.3 Conceptual framework

2.3.1 Baseball

Baseball became known as a competitive sport played with a hard ball and bat between two teams of nine players each on a field known as a diamond shape, with 3 base paths and a batting base.

It is impossible to be sure where the first baseball game was actually played in Mexico; several cities claim the honour and despite efforts to discover the exact location, historians still cannot agree. But looking at all the studies done, three cities come closest to qualifying: Guaymas in the state of Sonora, Nuevo Laredo in the state of Tamaulipas and Cadereyta Jiménez in the state of Nuevo León.

The Guaymas account is more accurate and states that it was in 1877 when the sailors who made up the crew of the American ship Montana, visiting Guaymas, went ashore and played a game of baseball with each other.

2.3.2 Baseball

A baseball is a ball used to play the sport and is easily recognisable by its characteristic red stitching. The ball has a sphere shape and a core of rubber or cork wrapped in thread. The core of the ball is covered by 2 pieces of white leather in the shape of peanuts sewn together by red thread. The stitching is an important element that influences the drag of the sport.

Regarding the particular shape of its seams, some researchers agree that they are so because when the sport began to be played in the 19th century, making them with that design was the only way to make a completely spherical ball. The standard for their aesthetics was set in the 19th century. The seams not only have to do with the design of the ball, but also with the way it is thrown.

2.3.3 Design Software

It is a programme that enables mechanical design processes to be carried out, from the definition of needs or the conception of the idea by the designer to the creation of the technical drawings necessary for its manufacture. By means of an interface where the programme and its part design, assembly and drawing tools are a working method for the design, the operator can model the part in three dimensions and quickly create the necessary views for the conception of drawings.

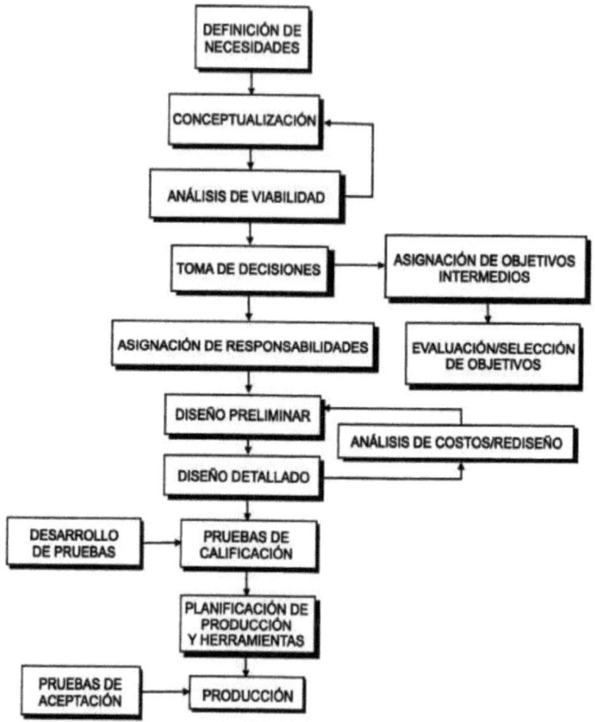

Figure 2 Stages in the design process, taken from the text "The Engineering Design.

2.3.4 Engine

The motor converts either electrical, chemical, or any other energy into a rotating mechanical motion.

Engines are classified by the type of energy they use, the work they are capable of doing in a unit of time at a given rated speed, which is the angular velocity of the crankshaft, i.e. the number of revolutions per minute (rpm or RPM).

The engines are used in applications such as:

- Kitchen appliances (rubbish processors, blenders, mixers , food processors).
- Sewing machines.
- Hoovers.
- Hand tools (saws, drills).

2.3.5 Coupling of an electric motor

The coupling of an electric motor is a device that connects the motor shaft to the equipment that the motor is to drive, i.e. the motor coupling allows the motor to act on the driven equipment.

The correct coupling for an application is selected by determining the rated torque of the power source, determining the service factor, calculating the coupling torque capacity and ensuring that the coupling is the correct size on the shaft to mate with the unit to be driven.

2.3.6 Slots

Precision location is very important in various engineering applications such as machining and assembly. The tool follows a very precise path and a workpiece must be accurately and stably located in a precise position. In assembly, the positions of the assembled parts must be easily assembled and excessive constraint of the parts must be avoided. One of the common techniques to achieve these goals is the use of slots as part features.

The use of two pins placed in two holes poses both location and manufacturing problems.

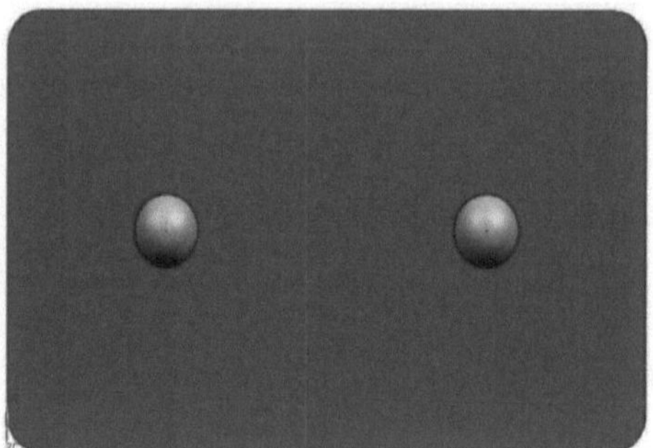

Figure 3 Part without slots

a) Problems with assembly

1. Impossible to determine the exact location of the workpieces.

2. Creation of high stresses in the workpiece as a result of excessive restriction of the workpiece location.

b) Problems with manufacturing

1. Holes in the workpiece must be machined to very tight position and diameter tolerances, which increases the cost of manufacture.

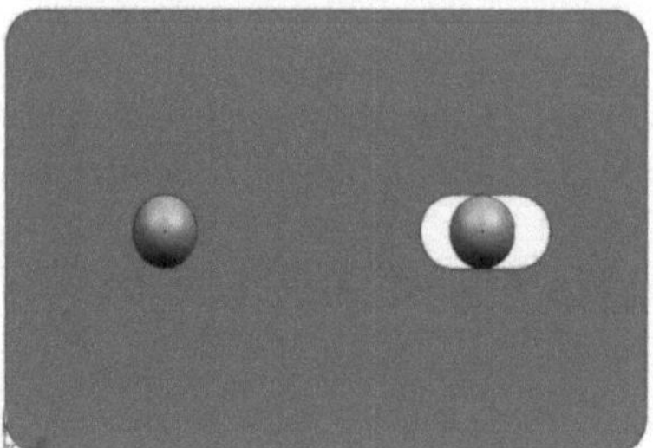

Figure 4 Slotted pin.

To overcome these problems, one of the pins can be assembled in a slot. Thus, the pin in the hole eliminates two degrees of translational freedom, while the one in the slot eliminates the last degree of rotational freedom.

In the following section, we will look at the use of slots to overcome the problem of over-constraint through a sample part using the principles of geometric dimensioning and tolerancing as defined by the ASME standard [1].

c) Use of slots: an example of GD&T

As we saw earlier, the slots allow us to avoid excessive restrictions and to have greater precision. An example of this can be seen in the illustration.

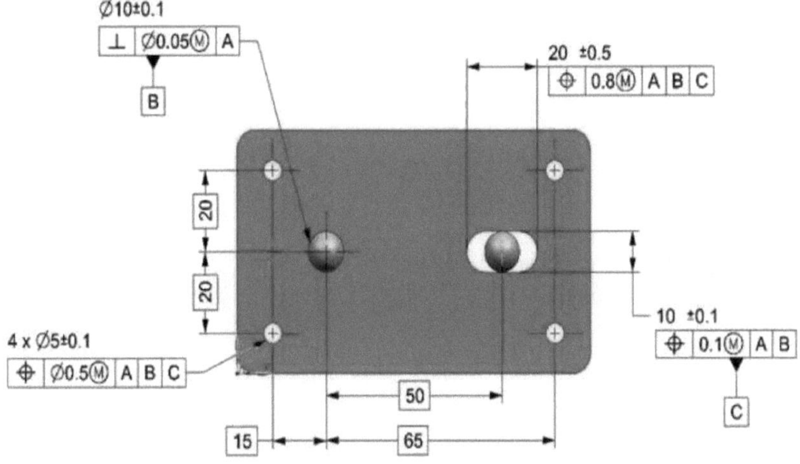

Figure 5 Artwork with slot tolerances applied.

From Figure 5, we can observe the following:

- Main reference function (A): the plane at the bottom of the part (not shown)
- Secondary reference feature (B): The defined bore with a feature size tolerance of ±0,1 mm and a perpendicularity tolerance of 0,05 mm in CMM to reference feature A.

- Tertiary reference feature (C): the width (vertical dimension) of the slot with a feature size tolerance of ±0,1 mm and a position tolerance of 0,1 mm in CMM with respect to the reference frame AB.

- The length (horizontal dimension of the slot): It has a feature size tolerance of ±0,5 mm and a position tolerance of 0,8 mm in CMM with respect to the ABC reference frame. Following the logic of the use of slots, this feature has the highest tolerances.

- The 4-hole pattern: The holes in the pattern have a feature size tolerance of ±0,1 mm and a position tolerance of 0,5 mm in CMM with respect to the reference frame ABC. The position of these holes is defined using basic dimensions with respect to the reference frame ABC.

- Despite the advantage of overcoming excessive restriction, with slots we can still have the problem of high manufacturing cost if your part is of considerable thickness. This is because making a slot is simply more complex than drilling a hole in relatively thick parts. One technique to have the best of both worlds is to use diamond pins. We will explore this technique in a future publication.

2.3.7 FEA analysis

Finite Element Analysis (FEA) is the modelling of products and systems in a virtual environment to find and solve potential structural or performance problems. FEA is the practical application of the finite element method (FEM).

It is a mathematical model to represent a graph of loads in finite elements including deformation and stress analysis, this analysis is of vital help to optimise elements with the possibility of failure.

CHAPTER III
METHODOLOGY

3.1 Design

The aim of this chapter is to achieve a design that adapts to the proposed requirements and functionalities of the machine in order to subsequently carry out a CAD/CAE modelling using SolidWorks software that allows the project to be sketched and the feasibility of positioning structures and components to be determined. The final design of the structure of the machine is then presented, which is based on the triggering of the machine by means of the propulsion system, a catapult type mechanism.

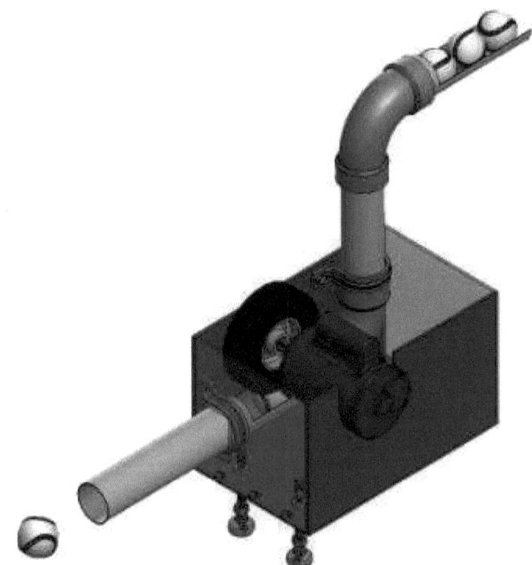

Figure 6 CAD design of ball throwing machine

3.2 Requirements

The requirements are a fundamental part of a conceptual design which proposes a direction and limitations for the development of the project, in this case, the following characteristics were taken as a reference:

- Size of a professional baseball.
- Approximate launch of 90 km/hr.
- Easy assembly and disassembly to replace damaged parts.
- Easy transport of the machine.
- Control launch angle.
- Total quote less than 14,000 Mexican pesos.

3.2.1 Size of a professional baseball

Professional baseballs have a diameter between 2.70"-2.81" (7.3-7.5 cm) and a circumference of 9"-9.25" (22.9-23.5 cm). The mass of a baseball is between 5-5.25 oz (142-149 g).

For the creation of the machine design, this variable was taken into account in the selection of the materials for the throwing arm.

Figure 7 Conventional Baseball

3.3 Selection of materials

The vast majority of technological advances achieved in modern society have relied on the discovery and development of engineering materials and manufacturing processes used to obtain them. An adequate selection of materials and processes guarantees the designers of mechanical parts the correct functioning (performance) of the designed components.

Table 2 PVC material selection

Material	Length	Largest Diameter	Smaller Diameter	Quantity
Underground piping of PVC	4ft	3 1/4in	3.11 in	1
Elbow connectors 90° long PVC		3 3/4in		2

After the cuts have been made, the materials are integrated with a simple assembly using only glue:

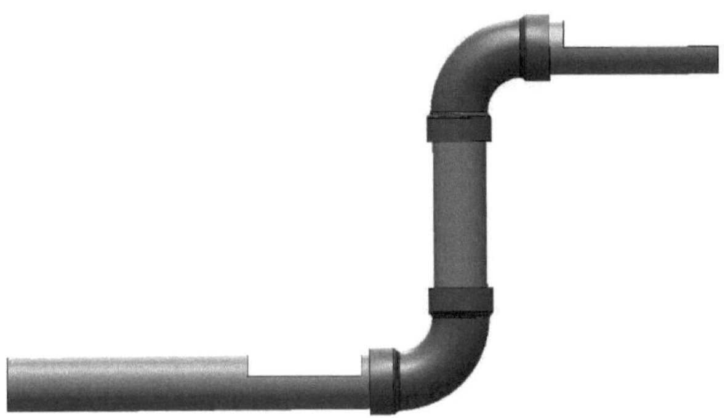

Figure 8 Sectioned pipe

3.4 Data collection

3.4.1 Approximate launch of 90KM/HR

For the selection of the motor, it is necessary to start from the speed we need to achieve, it is considered that the angular velocity of the motor is constant, under these terms, we can start using the theorem of uniform circular motion "MCU".

Some of the main characteristics of uniform circular motion are as follows:

The angular velocity is constant (ω = cte).

The velocity vector is tangent at each point on the trajectory and its direction is that of the movement. This implies that the motion has normal acceleration.

Both the angular acceleration (α) and the tangential acceleration (at) are zero, since the speed or celerity (modulus of the velocity vector).

Applying these concepts to the present project, we consider the desired speed:

$$V = 90\,\frac{km}{hr}$$

To work with this theorem it is necessary to convert to centimetres over seconds:

$$V = 90\frac{km}{\cancel{hr}}\,x\,\frac{1\cancel{hr}}{3600seg} = 0.025\frac{\cancel{km}}{seg}\,x\,\frac{100000cm}{1\cancel{km}} = 2500\,\frac{cm}{seg}$$

Applying these concepts to the present project, we consider the desired speed:

$$\text{Ecuación} \quad V = W x\, r \qquad (1)$$

$$W = \frac{V}{r}$$

Speed and radius variables are substituted

$$W = \frac{2500\frac{cm}{seg}}{10.16cm} = 246.062\frac{rad}{seg}$$

Conversion from rad over seconds to RPM is performed.

$$W = 246.062\frac{rad}{seg} \, x \, \frac{1\ vuelta}{2\pi rad} \, x \, \frac{60\ seg}{1\ min}$$

$$W = 2350.91\ RPM$$

To select an engine for this project, we have to consider an engine close to 2350 RPM and of an affordable low cost,

As a matter of cost, the selection of the engine was as shown in the following image:

Model No: 048A17T2006
Catalog No: X921
1/3,1725,TENV,48Z,1/60/115/230
Pedestal Fan

Figure 9 Engine 048A17T2006

Table 3 Parameters for approximate launch

Parameter	Variable
HP	1/3
Voltage	115/230
Kilo Watts	0.25
Amperes	3.8/1.9
Phases	1
Weight	12 lb
RPM	2250

To know the precision of the linear velocity at which the ball will be thrown with this motor, we again apply MCU:

Clearing θ from angular velocity

$$\text{Ecuación} \quad W = \frac{\theta}{t} \qquad (2)$$

$$\theta = 2250\ x\ 2\pi rad$$

$$\theta = 4500\pi rad$$

$$W = \frac{4500\pi rad}{1\ min} = \frac{4500\pi rad}{60\ seg} = \frac{4500\pi rad}{6\ seg} \qquad 4500\pi rqad$$

Substituted in linear velocity formula:

$$V = W x\, r$$

$$V = \frac{4500\pi rad}{6\, seg} x \frac{4\, in\, x\, 2.54cm}{1\, in} = \frac{4500\pi rad}{6\, seg} x\, 10.16cm = \frac{4500(3.14)\, cm}{6\, seg} = 2355 \frac{cm}{seg}$$

$$V = \frac{(2355 \frac{cm}{seg})\, x\, (3600 \frac{seg}{hr})}{10000cm} x1km$$

$$V = 84.78 \frac{km}{hr}$$

Figure 10 Engine sub-assembly.

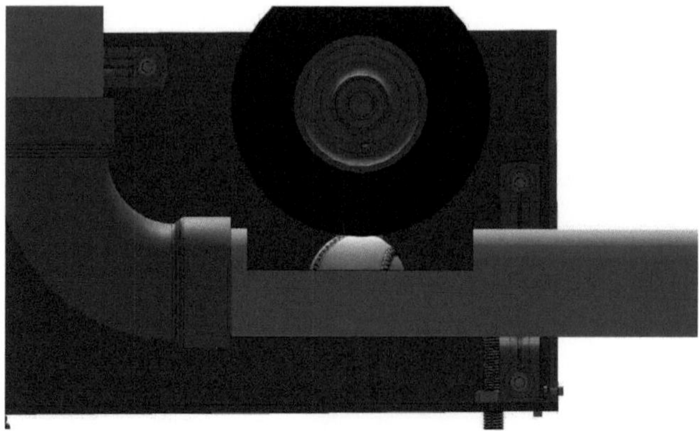

Figure 11 Launcher sub-assembly

3.4.2 Launch angle or departure angle control

Angle of exit of the ball: This refers to the moment of exit, at the point where the ball leaves the cue ball after it is released by the throwing hand, once it is thrown. This angle is between the perpendicular axis that passes through the body forming the (X) axis, and the (Y) axis that starts from the height of the shooter's face towards the front of the shooter. This can be positive or negative.

Figure 12 Exit Angle

- Positive: when the ball thrown by the pitcher is received by the catcher in the area determined by him before throwing it, or in its effect is directed to a certain area of the area marked as strike zone, which will have 9 parts of equal size and in which points will be given according to the area where the ball falls, the better the effectiveness, the better the result (there is presence of control).

- Negative: when the pitcher's throwing tends to uncoordinated movement, disorientation of the strike zone, and the pitched ball is received outside the zone agreed upon as the strike zone. (No control is present).

Wrist whip action at the moment of release of the ball: This refers to the final moment before the ball is released; it is the action of the wrist of the throwing hand.

Adding the 4 inch bolt-on swivel level allows the mechanism to have 0 to 11.71 degrees of inclination for parabolic shots, this level will be adjusted with a crescent spanner depending on the type of training to be performed.

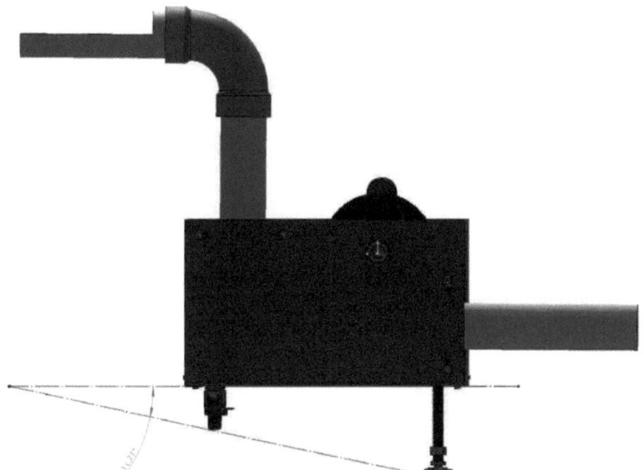

Figure 13 Top view inclination of the mechanism.

Figure 14 Front view of the mechanism

3.5 Assembly

The machine is designed to replace any damaged items with the exception of the welded structure, the only tools required to dismantle and assemble the machine are:

- Metric allen key ring
- crescent spanner 6 in

The following image shows an exploded view of the general assembly of all relevant items.

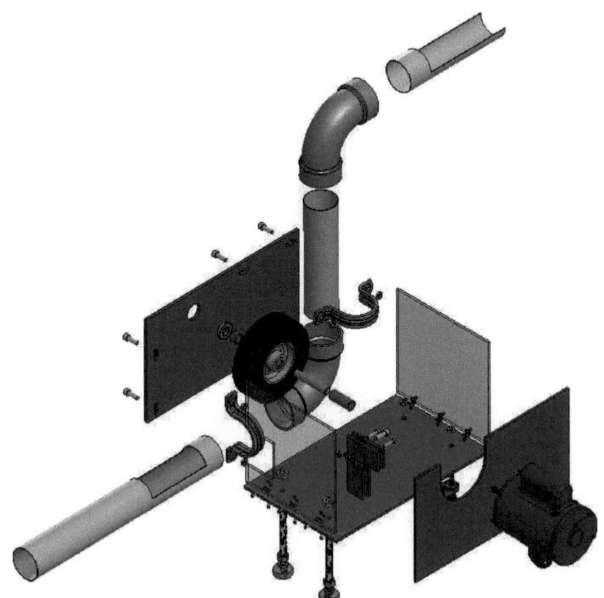

Figure 15 Exploded view of assembly

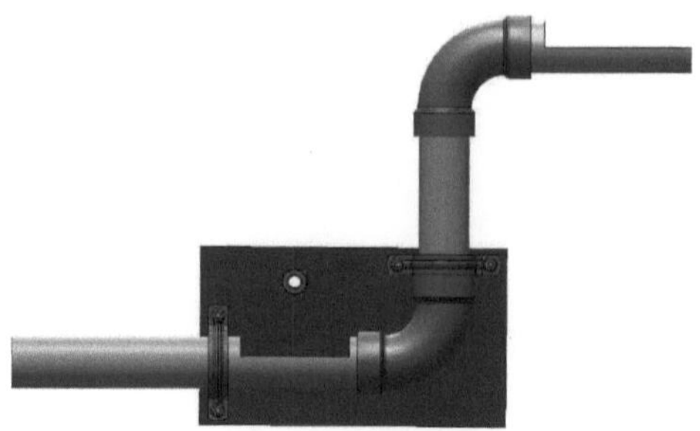

Figure 16 Throwing arm assembly with slotted plate

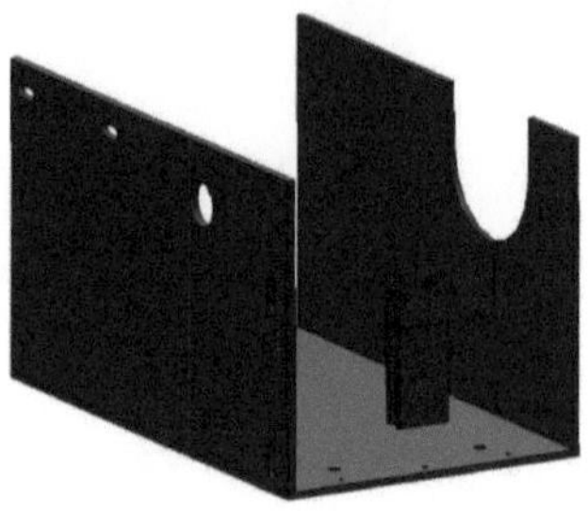

Figure 17 Welded plates.

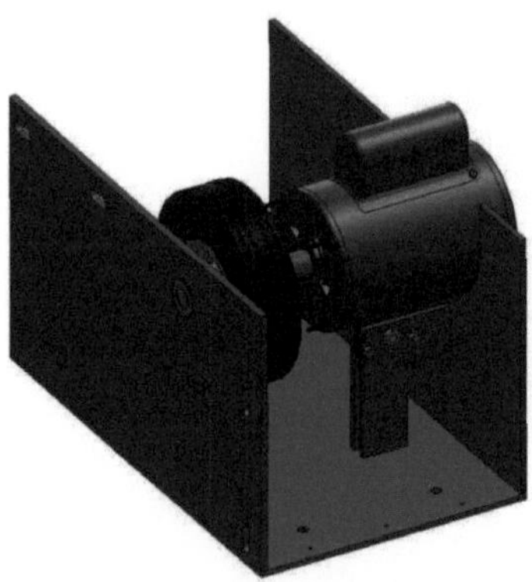

Figure 18 Isometric view engine assembly

For the safe and proper transport of the machine, standard rubber swivel castors with safety locks were chosen.

Figure 19 View of swivel castors on machine

3.6 Design optimisation

One of the problems to solve during the design of the machine was with the variable dimension of a ball, as we discussed in chapter 3.2.1, a professional baseball has a diameter between 2.70"-2.81", these dimensions are not exact depending on the brand of the manufacturer at the beginner level.

As a solution, 31mm slots were integrated, which displaced the PVC clamps.

The tangent measurement between the 8" wheel and the PVC bore surface is 2.65" when the bolts are at the midpoint of the slot.

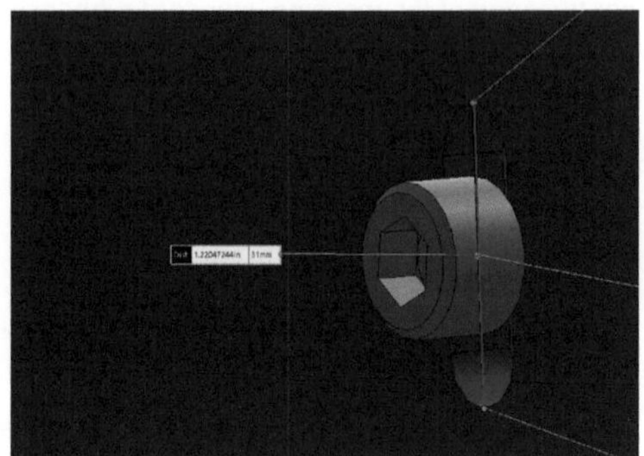

Figure 20 Distance of 1.22 in in the slot.

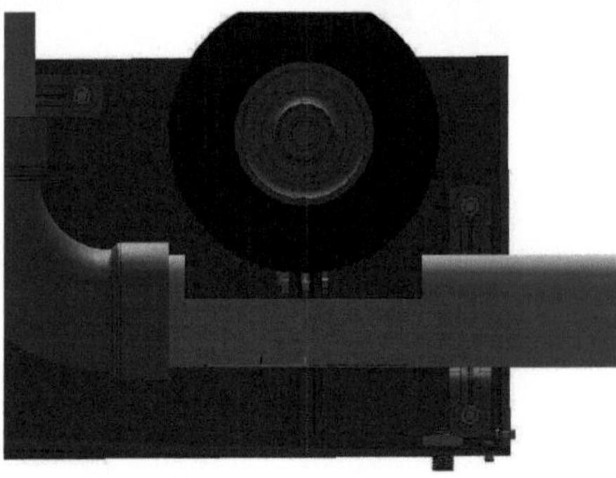

Figure 21 View of the distance between the wheel and the 2.66 in. launching arm.

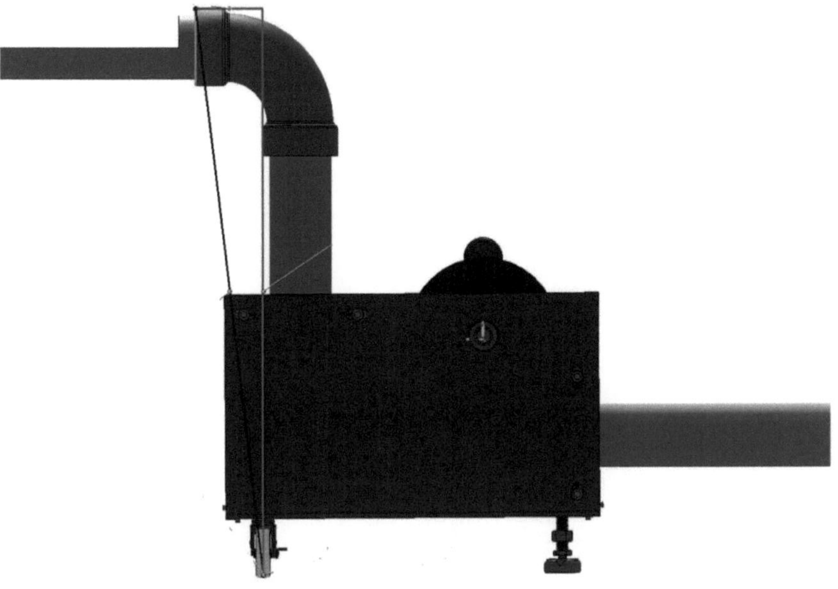

Figure 22 Maximum machine height 28.9 in.

Figure 23 Machine width 14.2 in.

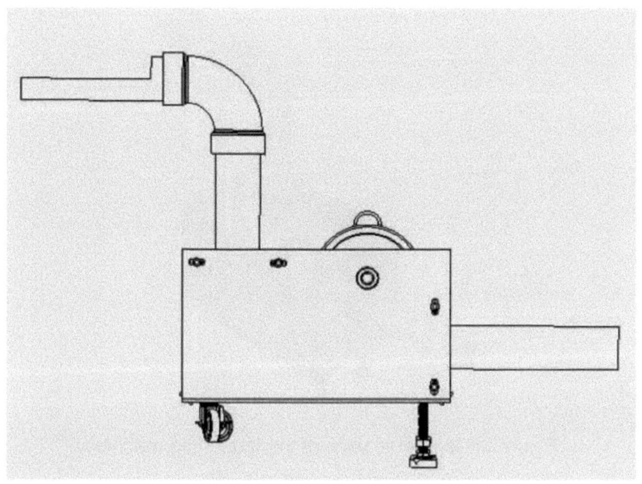

Figure 24 Machine length 43.8 in.

Table 4 Mechanical properties of the machine

Mass	25043.65 grams
Volume	10935467.97 cubic millimetres
Surface area	2774348.60 square millimetres
Centre of mass (millimetres)	X = - 55.50 Y = 101.26 Z = 79.35

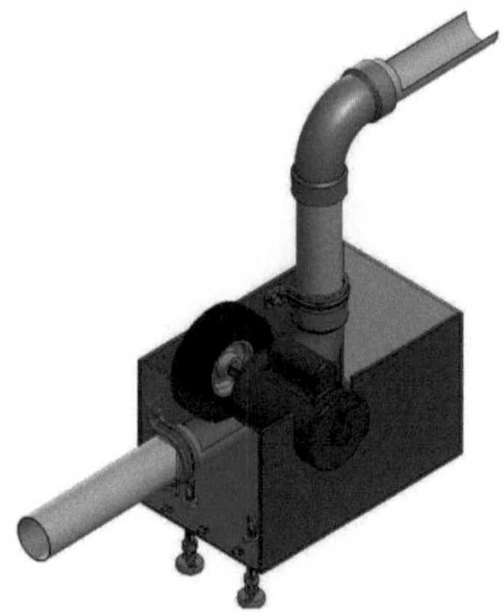

Figure 25 Isometric view of the launching machine.

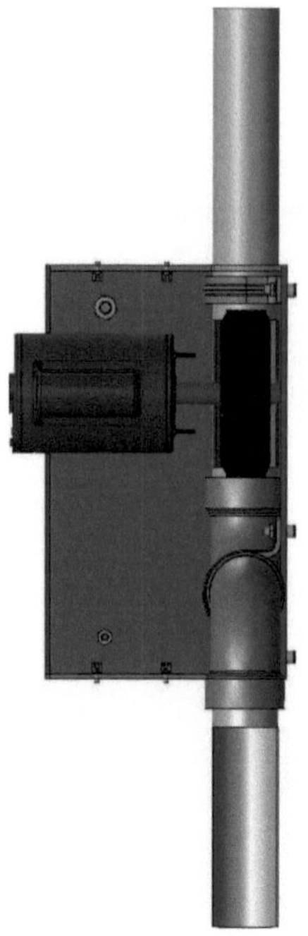

Figure 26 Plan view (top) of the launching machine.

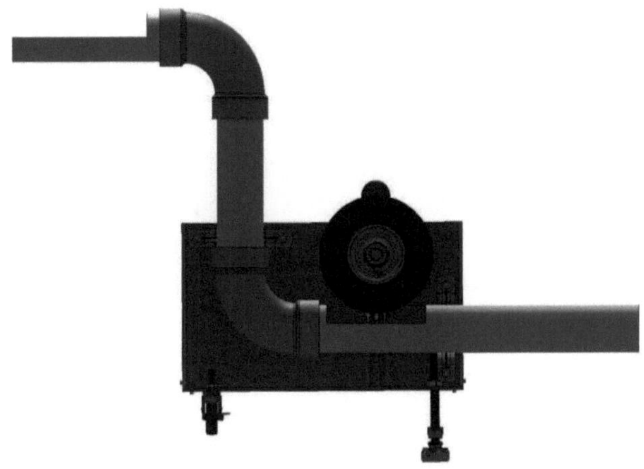

Figure 27 Top view (front).

3.7 FEA engine mount analysis

In the mechanical design of the ball machine, we selected a 12 lb = 5.45 kg motor that will be supported by a 2 x 3.9 x .25 inch carbon steel plate, in this plate sits the highest load of a component in our system, to verify that this element can withstand the vertical loads exerted in the holes, an FEA study is performed.

Figure 28 Load plate location supporting the weight of the engine.

To begin the study, we must define the material parameters, applied fasteners, load locations and a mathematical mesh.

In this project we selected the material carbon steel as cast carbon steel.

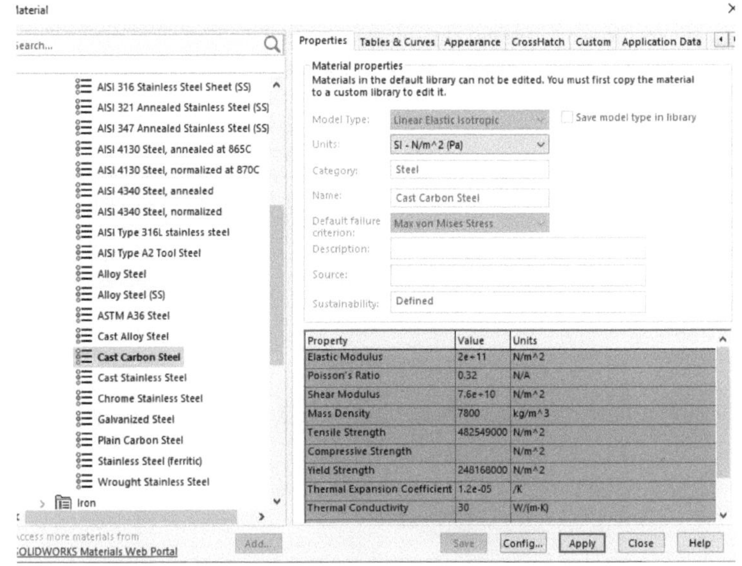

Figure 29 Mechanical properties of carbon steel.

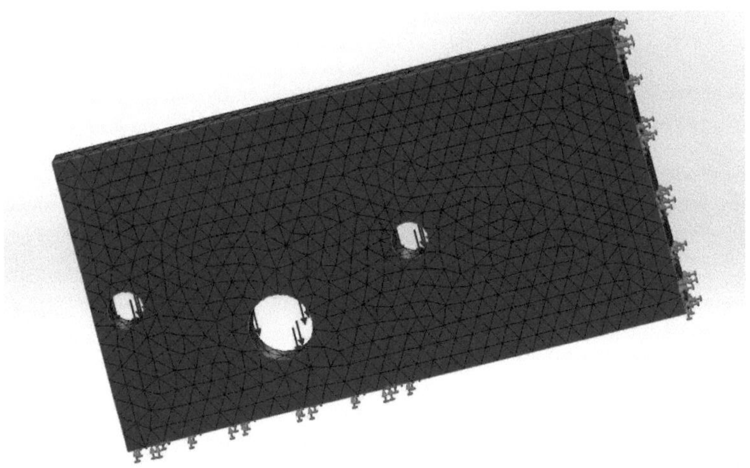

Figure 30 Location of loads, mountings and meshing on motor mounting plate

The fasteners are represented by green arrows, the loads are purple and the mesh is the triangular joints on the surface of the part, each joint is the mathematical representation in that position of the mathematical calculation of displacements and stresses generated by the SolidWorks program the component.

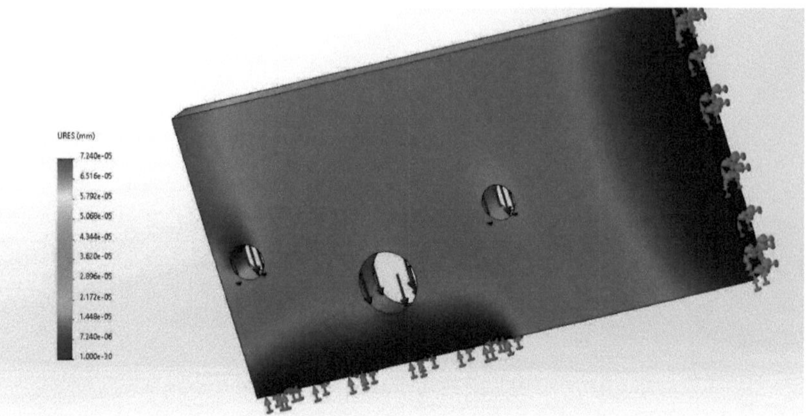

Figure 31 Displacement graph on motor clamp plate.

The result of the deformation study is represented by a graph on the left of figure 33, with a scale in millimetres shifted from 1x10^-30 in blue to 7.240x10^-05 in red.

The analysis indicates that the upper area of the first small hole has the largest deformation in our element with the total of 0.0000724 millimetres of deformation, which we conclude that its maximum deformation is negligible.

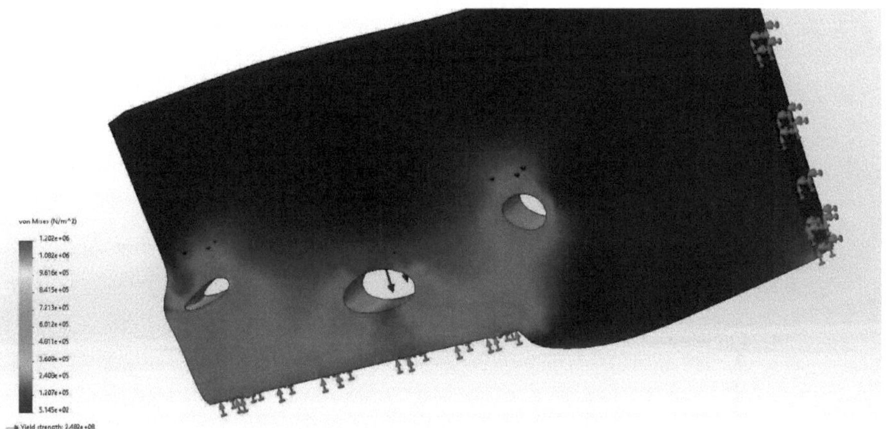

Figure 32 Von Mises Stress Graph

The stress study is represented in figure 34, where the simulation is shown in an exaggerated way to appreciate the behaviour of the part with the applied loads, in the same way, a Von Mises stress scale is found in N/m^2 , showing in blue colour a value of $5.145x10^\wedge\ 2\ N/m^2$ up to the maximum value represented in red with 1.202x10^06 N/m^2

The yield strength marked for the carbon steel material is 2.482x10^8 N/m^2 , we analyse that the maximum stress exerted on the part is much lower than the value of the yield strength, we conclude that the stress exerted on the engine mount will not fail due to engine loads.

CHAPTER IV
RESULTS

A ball throwing machine was designed with the measurements.

- 28.98 in maximum height
- 43.81 in Length
- 14.24 in wide

With a total weight of 25kg.

Launch speed at 84.78 km/hr, at a spin of 2250 RPM. Performing horizontal pitches from 0 to 11.71 degrees inclination.

A quotation was made through 5 different sites to obtain the materials used in this project, with price variations in each of the products and a final sum of all the cheapest products found in the market as shown ìtable 4.

With a total unit cost of 6,033.89 Mexican pesos per product, the sum total of each item for the quantity needed to be purchased gives a total of 6,934.18 Mexican pesos. In the market we found ball launching machines for up to 26,000 Mexican pesos, if we compare the cost of the present project with the market prices, we have a saving of 19,065.82 Mexican pesos.

The machine has the capacity to store 3 balls in the tank, area of improvement for the machine is in the tank to control time of passage per ball, this could be achieved with an arduino that can integrate an obstacle to the balls.

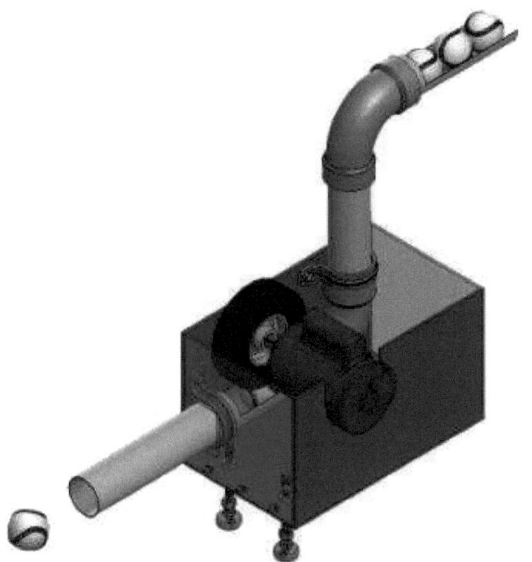

Figure 33 Isometric view of pitching machine with balls

Table 4 Material specifications and costs.

Material specification	Quantity	Mc Master	Free market	Google Shop	The home depot	Ebay	Selection
8in diameter wheel and 3/4in bore	1	$1,552.15	$950.00	$507.00		$573.00	$507.00
Engine 2250 RPM	1	$5,120.80	*	$2,731.00		$2,380.00	$2,380.00
Delrin 1 1/4 - 3/4in	1	$320.00	$420.00	$210.00			$210.00
Steel bar 3/4 x 12 in	1	$723.00	$510.00	$400.00		$153.00	$153.00
PVC connector for 3 in 90 degrees	2	$322.15	$115.00	$172.92	$28 .0	*	$28.00
PVC DI = 3in OD= 3.11in x 12in	2	$428.09	$233.00		$120.00	*	$120.00
Clamp omega 3 1/2 in	2	$133.20	$72.00	$93.83	$12.00	*	$12.00
Stainless steel plate carbon 24x24 in	1	$2,451.25	$2,300.00	*		$640.00	$640.00
Ball bearing R12 3/4" 1 5/8" 5/16	1	$797.91	$175.00	$175.00			$175.00
Laxan of 12x12x1/4	2	$1,200.00	$850.00	$850.00		$939.00	$850.00
Aluminium angle bracket 6 x 7/16 x 1/2	1	$1,300.00	$1,278.80	$629.00	$800.00		$629.00
Socket M4x30	12	$12.95			$3.70		$3.70
Socket M4x12	5	$12.95			$3.70		$3.70
M6 thread	2	$9.25			$2.00		$2.00
M4 thread	12	$9.25			$2.00		$2.00

Socket M6x10	2	$9.25			$3.70		$3.70
Rotatable threaded rod, drainer	2	$183.89					$183.89
Bolt-on swivel levelling	2	$534.65	$122.00				$122.00
M10 thread	5	$14.80			$3.50		$3.50
Thread 5/8	2	$9.25			$2.70		$2.70
Thread 3/8	2	$9.25			$2.70		$2.70
Total cost						**$ 6,934.18**	

CHAPTER V
CONCLUSIONS

The project presented shows a benefit in the cost of materials for the construction of a ball launching machine, and is able to compete with the brands on the market by omitting the manufacturing processes.

The machine has a weight of 25 kg, which makes it difficult to handle and transport. There is room for improvement for this project by reducing the thickness of the material used and reducing empty spaces inside the machine.

There are several configurations of launching machines that can be classified according to their propulsion system, degrees of freedom and number of rotary actuators. Each of these configurations allows the machine to fire at high speeds and to have autonomy from the user. However, each configuration has advantages and disadvantages related to the efficiency of the shot, which is why the propulsion system with two rotating rollers is the configuration with the best construction and cost performance, offering a higher rate of fire over long distances, provided that the weight of the rollers coupled to the motors is taken into account, in order to implement speed control more easily and a simpler mechanical design.

CHAPTER VI
RECOMMENDATIONS

As a starting point in the design of the ball throwing machine, consider the target speed parameter to be achieved, selecting the motor and continue with the structural design of the machine and with the gear, mould the design depending on the requirements given for its elaboration.

Material selection is extremely important in the planned design, always use as little material as possible and remove unnecessary space and material.

In the manufacture of fixtures we must also consider the manufacturing processes to which the material must be subjected, making cuts to the material with conventional tools preferably so as not to add to the final cost of manufacture, this should be considered during the design of the machinery to be manufactured.

When working with design software we must consider the capacity in our work team, the assembly of components can be quite complicated when the job has numerous elements to be assembled.

In addition, always use standard materials and measurements, as failure to work in this way will result in a complicated project for material sourcing and fabrication.

SOURCES

Harper, G. E. (2002). The ABC of electronic machine control.

Álvarez Lorente, Manuel and López Labat, Hermenegildo (2005). "Preparation and adaptation of the baseball pitcher's arm". Magazine

Álvarez Lorente, Manuel et al. (2002): "La efectividad del lanzador. Un retodel pitcheo contemporáneo". Digital magazine http://www.efdeportes.com/ Revista Digital - Buenos Aires - Year 8 - N° 45 - February 2002.

4.- ÁLVARO GONZÁLEZ, H., & HERNÁN MESA G, D. (n. d.). THE IMPORTANCE OF METHOD IN THE SELECTION OF MATERIALS. Scientia et Technica.

ASME Y 14.5-2009, Dimensioning and Tolerancing. New York: American Society of Mechanical Engineers.

Baseball Dimensions & Drawings | Dimensions.com digital http://www.efdeportes.com - Buenos Aires - Year 10 - N° 89 - October 2005. digital http://www.efdeportes.com - Buenos Aires - Year 11 - N° 106 - March 2007. 6.- Mechanical design with Solidworks 2015. (n.d.). Google

Books.https://books.google.com.mx/books?id=_o2fDwAAQBAJ

7.- Ealo De La Herrán, J. (2005). Béisbol. Ciudad de La H a b a n a : Editorial Pueblo y Educación; Third Edition.

8.- Empresarial, C. (2022, 31 July). The history of baseball in Mexico. Caribe Empresarial. https://caribempresarial.com/la-historia-del-beisbol-en-mexico/

9.- Garcia Melo, J. I. (2004). Fundamentos del Diseño Mecánico (1.a ed.).Universidad del Valle. https://books.google.com.mx/books?id=2RqUgt9YISEC&printsec=frontcover& dq=what+is+the+mechanical+design&hl=en&sa=X&redir_esc=y#v=onepa ge&q&f=false

10.-González García, Iván and Hernández Mayán, René (2007): Béisbol: algunas consideraciones sobre los lanzadores. Journal: la-habilidad-de-lanzar-de-los-os- pitcher-de-beisbol.htm https://books.google.com.mx/books?id=2Jp7EpxiMbwC&pg=PA150& dq=definicion+de+motor+electrico&hl=en&sa=X&ved=2ahUKEwi2-

57W9sr8AhXgJEQIHUEZCYYQ6AF6BAgCEAI#v=onepage&q&f=false

11.- Mexicano, E. P. H. B. (2019c, October 25). El Tec, 55 years of professionalism.El Heraldo de Juárez | Noticias Locales, Policiacas, sobre México, Chihuahuay el Mundo.https://www.elheraldodejuarez.com.mx/local/el-tec-55-anos-de- professionalismo-4365279.html

TOVAR, E. (Ed.) (2021). The benefits of machine tools-multitasking.

Modern Machine Shop, Modern Machine Shop Mexico. https://www.mms-mexico.com/articulos/los-beneficios-de-las-maquinas- tool-multitasking

Villalobos Trujillo, J. R., & Unzué Reina, A. (2008, August). System of exercises to improve control in the pitching skill of baseball pitchers. Efdeportes.com. Retrieved November 10, 2022, from https://www.efdeportes.com/efd123/ejercicios-para-mejorar-el-control-en-

ANNEXES

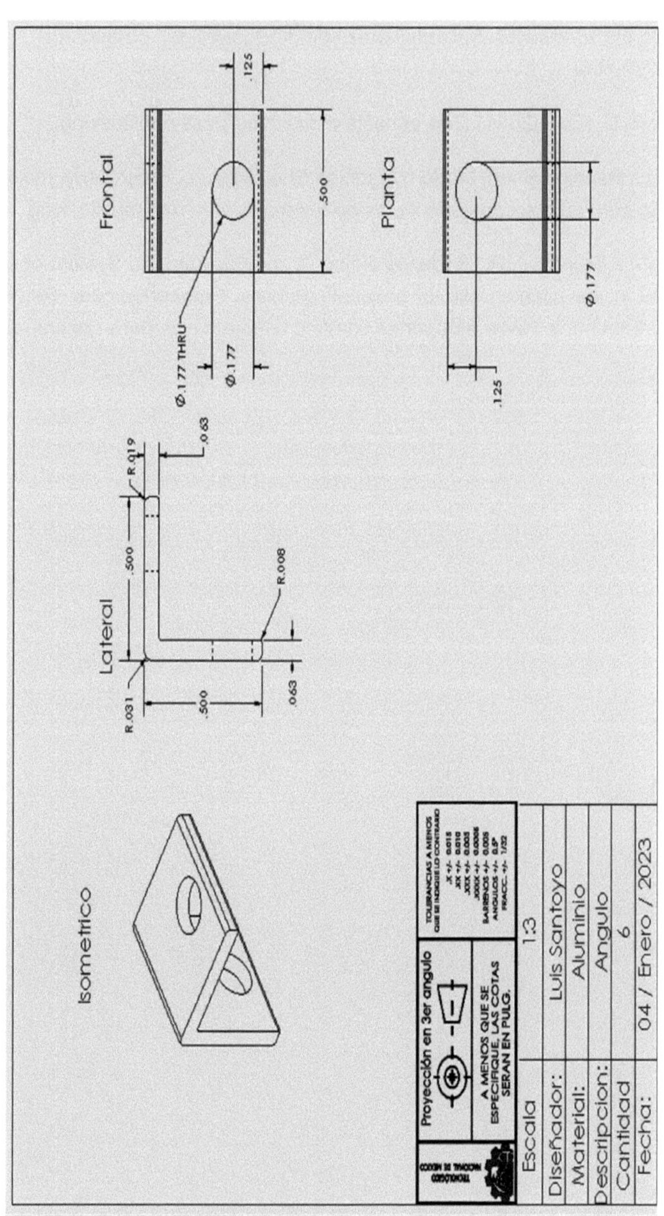

Figure 34 Plane angle.

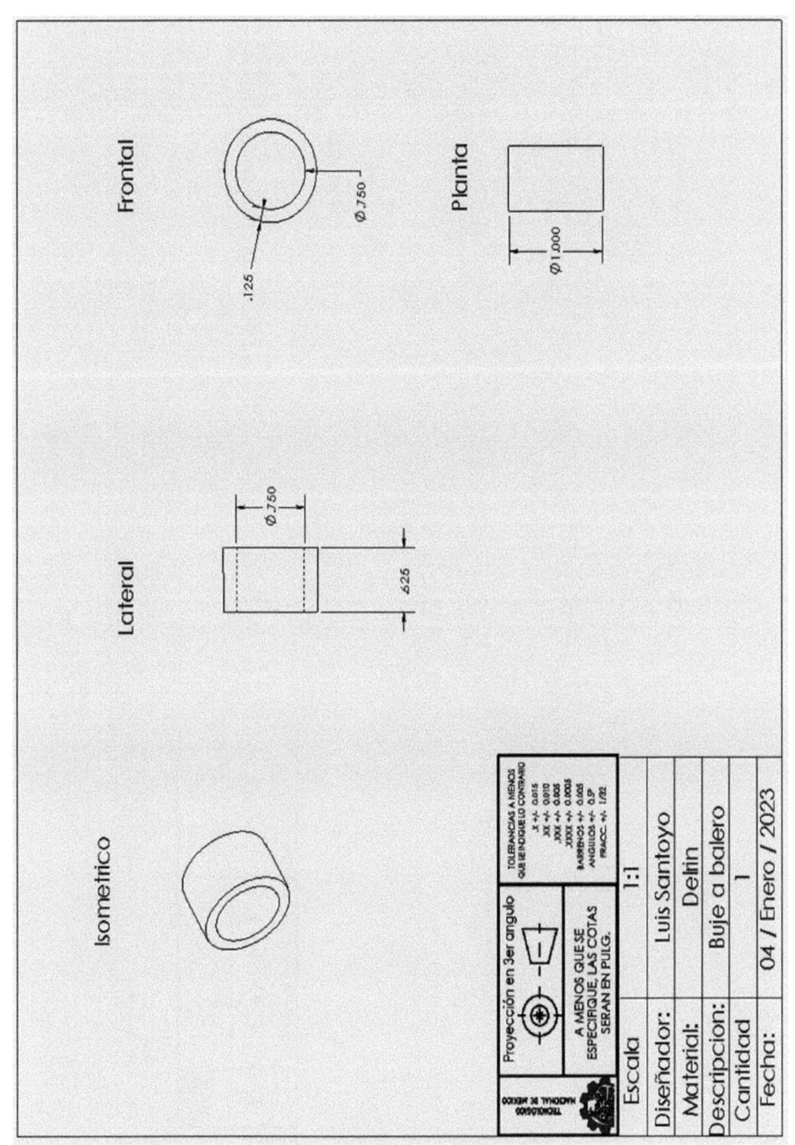

Figure 35 Bushing-to-bearing plane

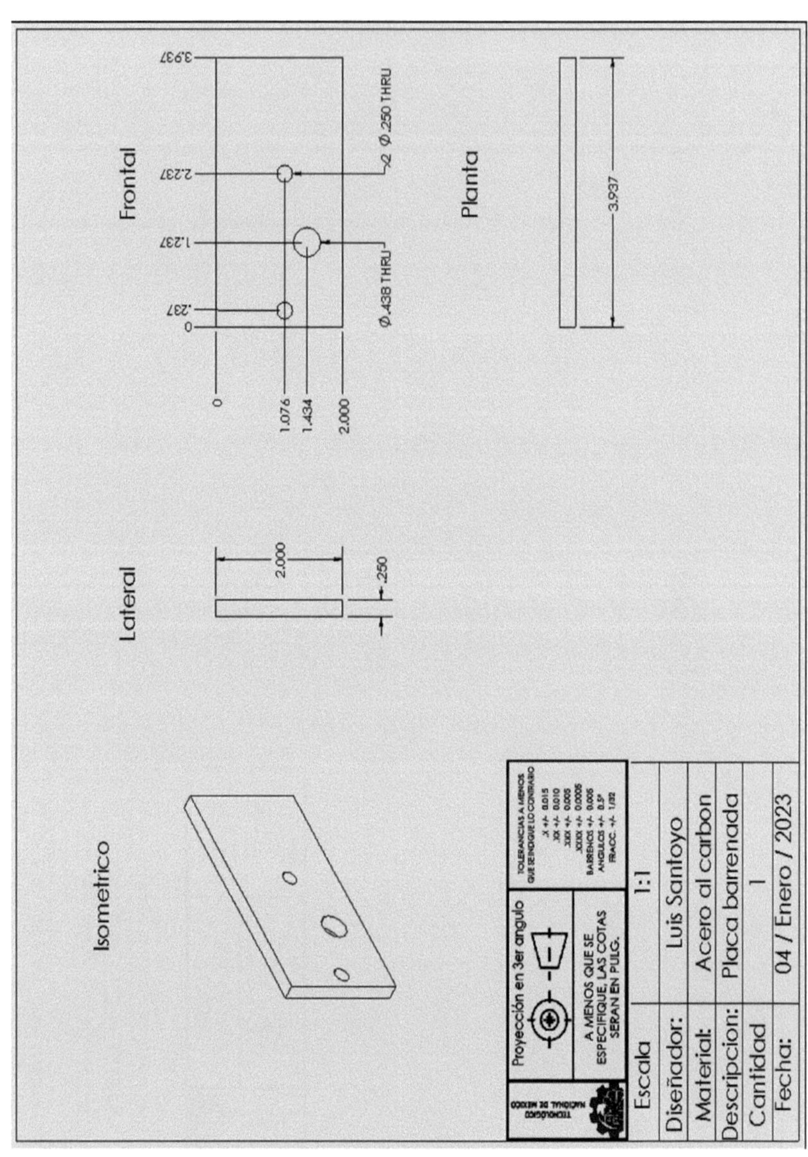

Figure 36 Bored plate

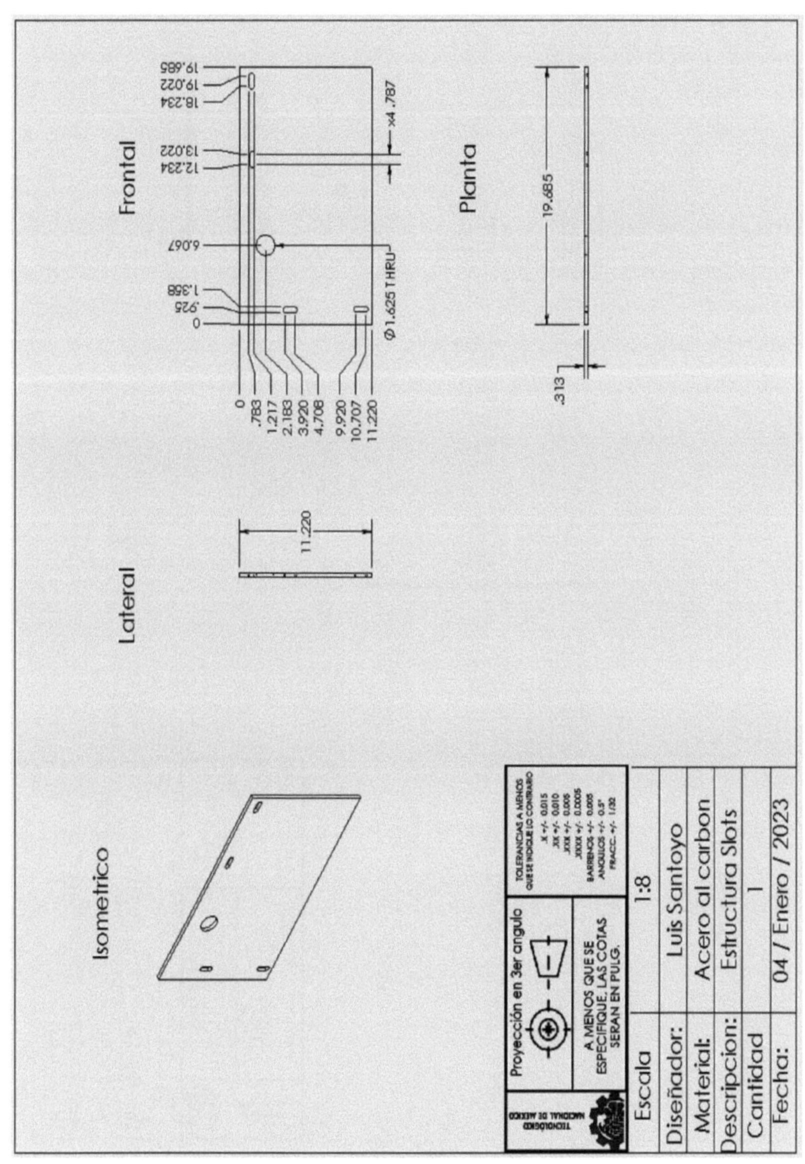

Figure 47 Slot structure plan

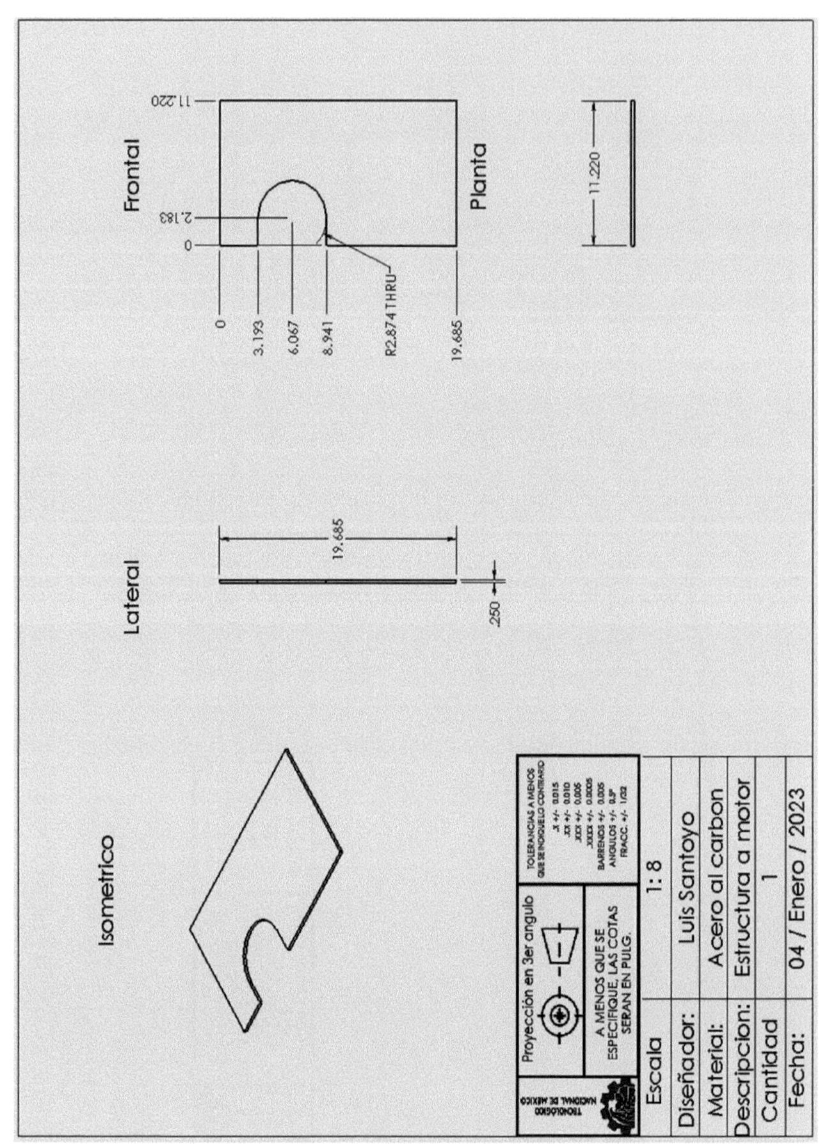

Figure 38 Motorised structure plan

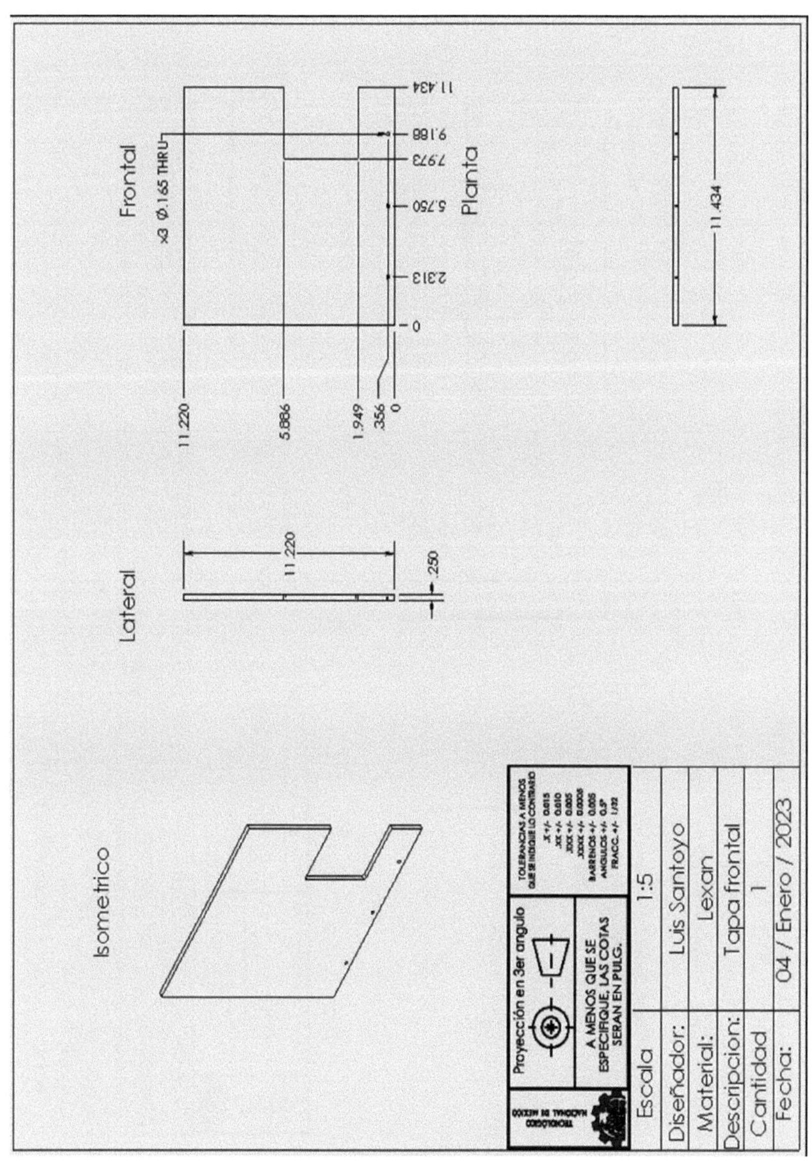

Figure 39 Front cover plan

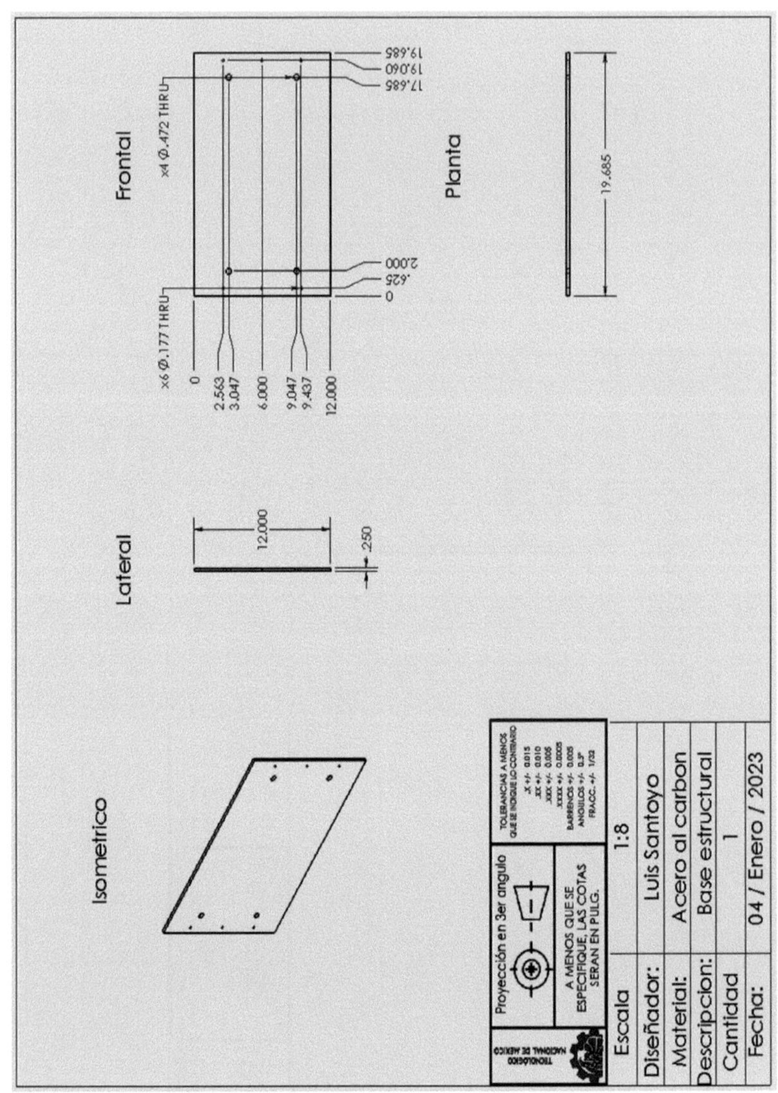

Figure 40 Structural base plan.

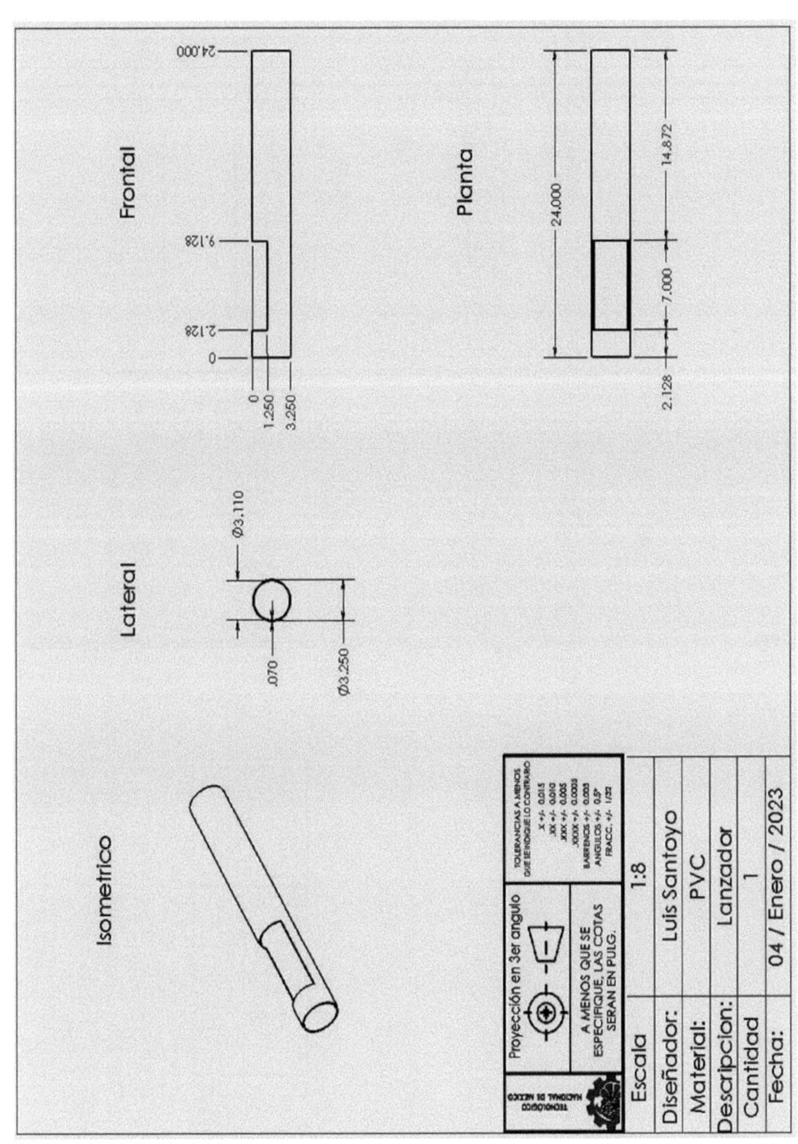

Figure 41 Launcher plan.

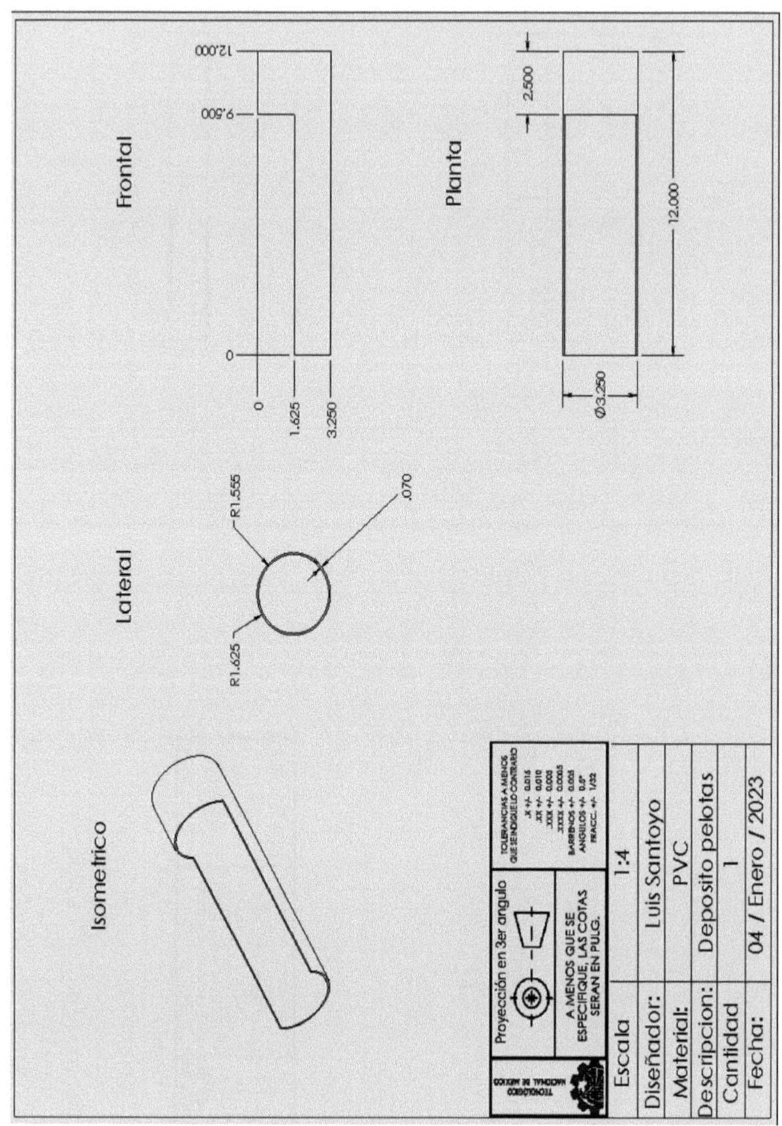

Figure 42 Ball storage plan.

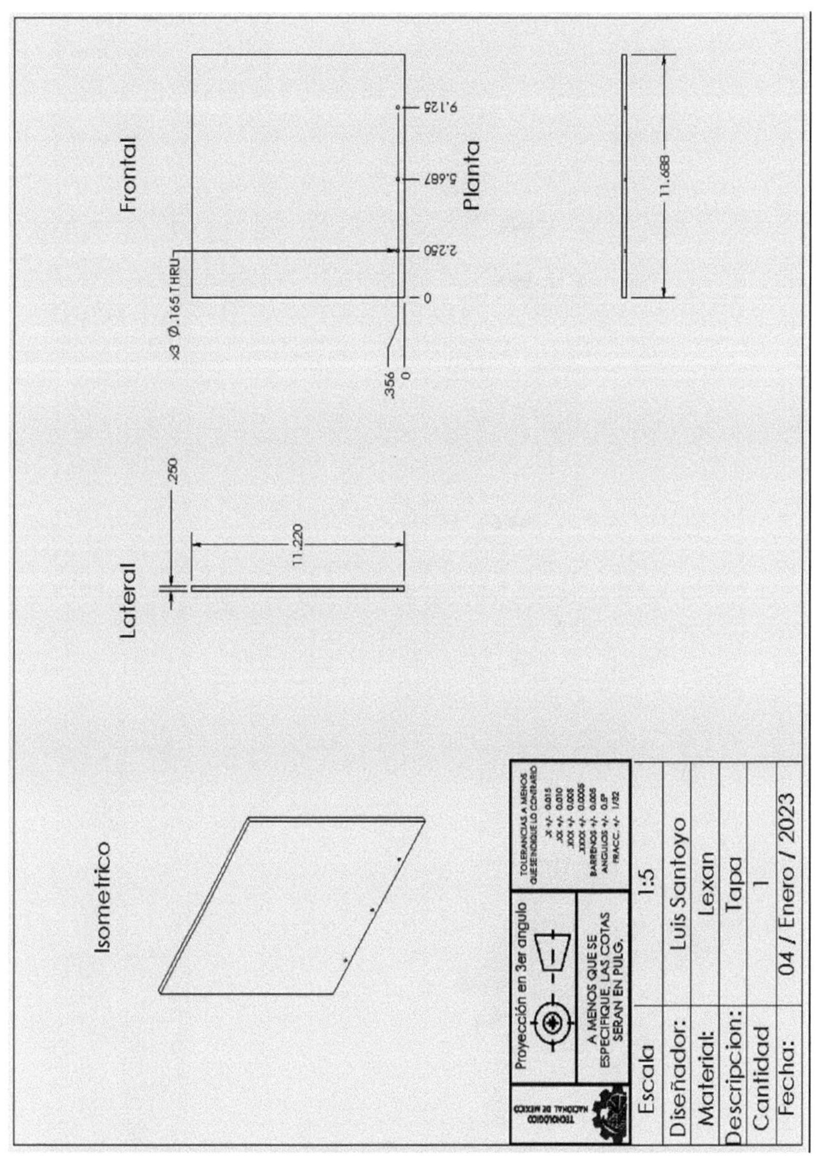

Figure 43 Cover plan.

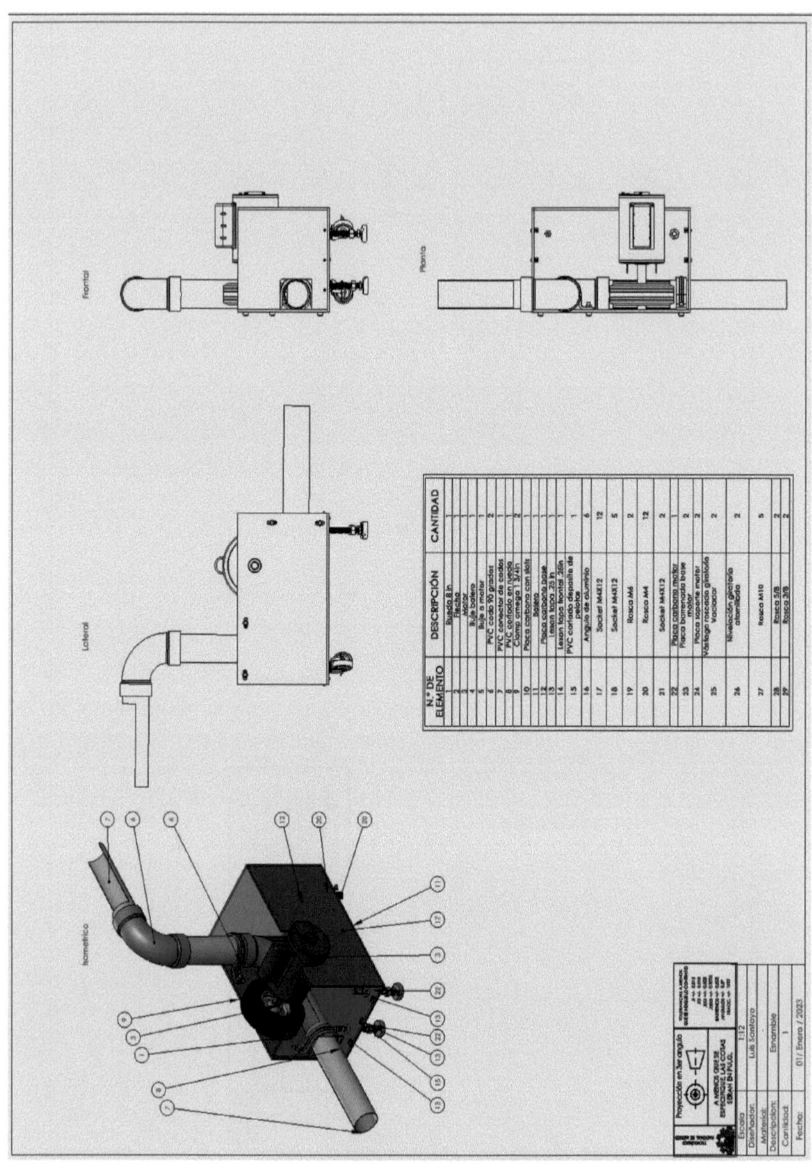

Figure 44 Assembly drawing.

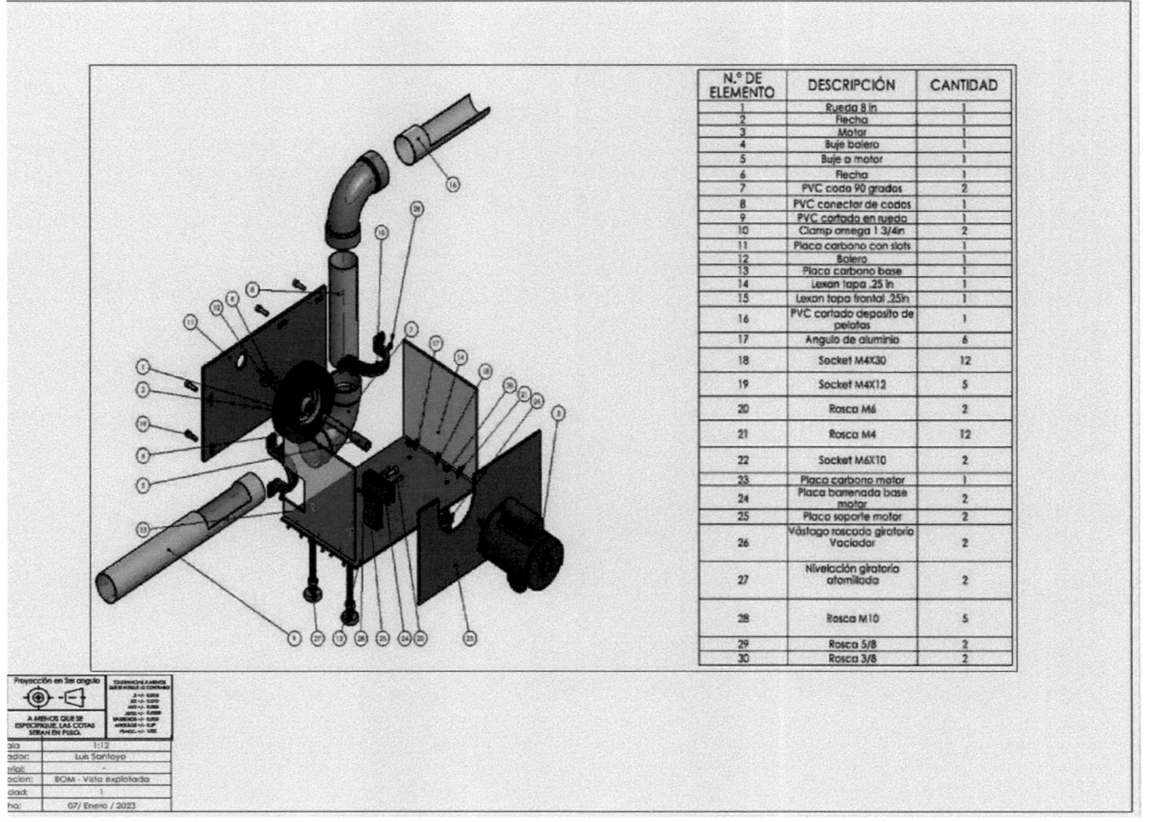

N.° DE ELEMENTO	DESCRIPCIÓN	CANTIDAD
1	Rueda 8 in	1
2	Flecha	1
3	Motor	1
4	Buje balero	1
5	Buje a motor	1
6	Flecha	1
7	PVC codo 90 grados	2
8	PVC conector de codos	1
9	PVC cortado en rueda	1
10	Clamp omega 1 3/4in	2
11	Placa carbono con slots	1
12	Balero	1
13	Placa carbono base	1
14	Lexan tapa .25 in	1
15	Lexan tapa frontal .25in	1
16	PVC cortado deposito de pelotas	1
17	Angulo de aluminio	6
18	Socket M4X30	12
19	Socket M4X12	5
20	Rosca M6	2
21	Rosca M4	12
22	Socket M6X10	2
23	Placa carbono motor	1
24	Placa barrenada base motor	2
25	Placa soporte motor	2
26	Vástago roscado giratorio Vaciador	2
27	Nivelación giratoria atornillada	2
28	Rosca M10	5
29	Rosca 5/8	2
30	Rosca 3/8	2

Figure 45 BOM of materials and exploded view.

yes

I want morebooks!

Buy your books fast and straightforward online - at one of world's fastest growing online book stores! Environmentally sound due to Print-on-Demand technologies.

Buy your books online at
www.morebooks.shop

Kaufen Sie Ihre Bücher schnell und unkompliziert online – auf einer der am schnellsten wachsenden Buchhandelsplattformen weltweit! Dank Print-On-Demand umwelt- und ressourcenschonend produzi ert.

Bücher schneller online kaufen
www.morebooks.shop

info@omniscriptum.com
www.omniscriptum.com

Printed by Books on Demand GmbH, Norderstedt / Germany